# A Street Called

## PLACE VALUE

By Toni Kay Porter

There once was a house,
an empty house,
on a street called, "Place
Value".

Three lonely numbers
with no last name,
moved into this house,
this empty house,
on a street called "Place Value".

Each number wanted a special
place,
a place they could call their
own.

They named their places,
Their very own places,
with special names – –
VERY special names,

"hundreds"    "tens"    "ones"

Sometimes all three numbers
were home,
and sometimes – they were
not...

When the number was home,
standing in his place,
his own special place,
that number had VALUE!

The number "7" was home,
standing in his place.
he looked down at his place,
looked up again – and smiled.
He stood proud and tall
and clearly said,
"My name is 7, and my value is 7 ones!"

His sister "4" came home
and stood in her place,
her very own place.
She looked down at her place,
stood proud and tall,
giggled and said, "My name is 4 and
my value is 4 "tens".

Four and seven held hands
and in unison said,
"Together we are 47!"
(Forty-Seven)

The door opened wide,
and in came "9".
He hopped into his place,
his own special place,
And called out,
"My name is "9"
and my value is 9 "hundreds"!

The three numbers joined hands
Standing tall and proud
And together they said,
"We are "947".
(Nine hundred forty-seven)

We are no longer lonely,
our house is not empty.
We know our place,
our own special place,
And we each have <u>VALUE!</u>

The End.

Now it's time for you to have some fun!

1. Make your own "Place Value" house!

2. On a piece of Manila paper, draw and decorate your own Place Value House.

3. Be sure to copy the place values, and put them in their "proper places"!

4. Copy and cut out the numbers on the following page.

5. Working with a friend, take turns mixing up the numbers and putting them in the Place Value House.

6. Tell your friend what the number is and also its value.

1     2     3     4     5

6     7     8     9

1     2     3     4

5 6 7 8

9 1 2 3

4 5 6 7

8 9

1     2     3   4     5

6     7     8     9

1     2     3     4

5     6     7     8

0 1 2 3

4 5 6 7

8 9

1     2     3     4     5

6     7     8     9

1     2     3     4

5     6     7     8

0 1 2 3

4 5 6 7

8 9

www.ingramcontent.com/pod-product-compliance
Lightning Source LLC
Chambersburg PA
CBHW041309180526
45172CB00003B/1037